The Little Book

of

Scientific Writing

Nancy Fox

2010
New Street Communications, LLC
Wickford, RI

newstreetcommunications.com

"Science, like life, feeds on its own decay. New facts burst old rules; then newly divined conceptions bind old and new together into a reconciling law."

- William James, *The Will to Believe and Other Essays in Popular Philosophy*, 1910

PHILOSOPHIÆ

NATURALIS

PRINCIPIA

MATHEMATICA.

Autore *J S. NEWTON,* *Trin. Coll. Cantab.* *Soc.* Matheseos
Professore *Lucasiano,* & Societatis Regalis Sodali.

IMPRIMATUR·
S. P E P Y S, *Reg. Soc.* P R Æ S E S.
Julii 5. 1686.

LONDINI,

Jussu *Societatis Regiæ* ac Typis *Josephi Streater.* Prostat apud
plures Bibliopolas. *Anno* MDCLXXXVII.

CONTENTS

Fig. 1.

A
B
C
B
E
D

SOME BRIEF
WORDS OF WISDOM

"Let thy speech be short, comprehending much in a few words."
- *Apocrypha*

"When you wish to instruct, be brief; that men's minds take in quickly what you say, learn its lesson, and retain it faithfully. Every word that is unnecessary only pours over the side of a brimming mind."
- *Cicero*

"The letter I have written today is longer than usual because I lacked the time to make it shorter."
- *Blaise Pascal*

"Simplicity is the ultimate sophistication."
- *Leonardo da Vinci*

"It is impossible to disassociate language from science ... To call forth a concept, a word is needed."
- *Antoine Lavoisier*

ON

THE ORIGIN OF SPECIES

BY MEANS OF NATURAL SELECTION,

OR THE

PRESERVATION OF FAVOURED RACES IN THE STRUGGLE
FOR LIFE.

By CHARLES DARWIN, M.A.,

FELLOW OF THE ROYAL, GEOLOGICAL, LINNÆAN, ETC., SOCIETIES;
AUTHOR OF ' JOURNAL OF RESEARCHES DURING H. M. S. BEAGLE'S VOYAGE
ROUND THE WORLD.'

LONDON:

JOHN MURRAY, ALBEMARLE STREET.

1859.

The right of Translation is reserved.

FOREWORD

Scientific research, theorems and results have no value or relevance whatsoever until they are communicated to peers. This communication must be executed in a clear manner which - though incorporating all relevant details by which colleagues can reproduce, assess and test reported results - nevertheless remains as concise as possible.

Publication is the face by which scientific researchers in any field - engineering, chemistry, physics, medicine, etc. - become known within their discipline. Thus all scientific researchers must train themselves in the subtle art of presenting their objectives, experiments and results both efficiently and elegantly.

One's science may be superb, but if one's writing does not deliver coherence and structure to match, then the message can too often be killed by the medium. Poor writing will prevent - or in the best-case scenario, delay - the recognition of good science.

One need not become Charles Dickens. Quite the contrary. Grandiosity and embellishment are useless to the scientist. One needs, instead, to understand precision, coherence and directness. One needs to learn how to organize thoughts in a logical, building-block manner. And one needs to understand how to map one's experiments with totally-focused clarity: achieving directness while at the same time leaving absolutely no step undocumented for colleagues who will seek to reproduce reported results.

This book - this *brief* book, which I hope is itself a model of the elegant compactness to be striven for in all writing - will show you how.

- Nancy Fox

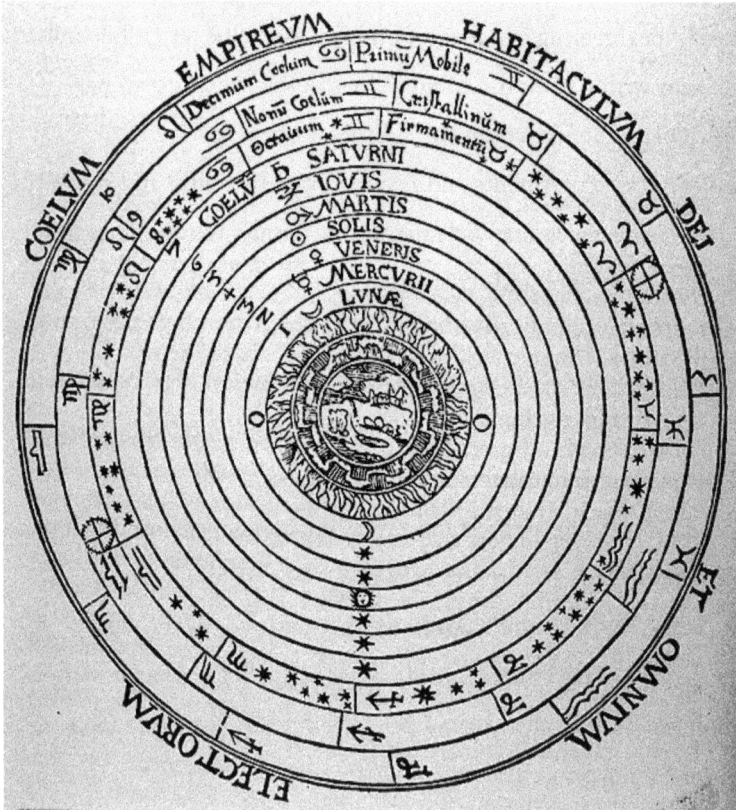

What Constitutes Valid Publication?

"If you cannot - in the long run - tell everyone what you have
been doing, your doing has been worthless."
- Erwin Schrodinger (Nobel Prize winner in physics)

A valid primary scientific publication is *the first disclosure of
new research results* in a forum that allows scientific peers to
study, test and fully comprehend the author's observations,
methodology, and conclusions. Under this definition, conference
reports, theses, abstracts and such are *not* considered valid
publication.

The forum for publication must be permanent, peer-
reviewed, available to the broad scientific community with no
restrictions, and archived through standard scholarly information
retrieval services (Chemical Abstracts, Index Medicus, and so
forth). Note that government reports, corporate bulletins and
other such mediums do not qualify.

ON OUR KNOWLEDGE

OF THE

CAUSES OF THE PHENOMENA

OF

ORGANIC NATURE.

BEING

SIX LECTURES TO WORKING MEN,

DELIVERED AT THE

MUSEUM OF PRACTICAL GEOLOGY.

BY

PROFESSOR HUXLEY, F.R.S.

LONDON:
ROBERT HARDWICKE, 192, PICCADILLY.
1863.

THE RITUAL OF THE SCIENTIFIC PAPER: IMRAD

"But words are things, and a small drop of ink, falling like
dew upon a thought, produces that which makes thousands,
perhaps millions, think."

- Lord Byron

Some things in this world never change: the rites and
incantations of various faiths, for example, or the rules of the
game of golf. The same is so with the rubric of the scientific
research paper, the rudimentary format of which has evolved
over generations and has by now become fundamental. This
rubric is referred to as the IMRAD format, codified by the
American National Standards Institute (ANSI).

As may be inferred from the acronym, the IMRAD format
consists of four sequential parts: *Introduction, Methods, Results,*
and *Discussion.*

Your *Introduction* will explain the question to be studied: the
problem you are attempting to solve. Your presentation of
Methods will explain the mechanics of the experiment you have
run in order to answer the question at hand. Your *Results,* of
course, are the findings of your experiment. Finally, your

Discussion will speculate on what the findings mean and what new questions the findings give rise to.

In detail, therefore, your paper will be comprised of nine sections: Title, Authors/Affiliations, Abstract, Introduction, Methods, Results, Discussion, Acknowledgments and Literature Cited.

The above formula - combined with the "Instructions for Authors" from the journal to which you will direct your paper - is the key to structuring your paper.

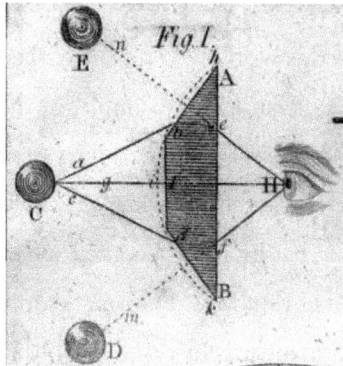

Fig. 1.

YOUR SHOUT-OUT: THE IMPORTANCE OF TITLE

"English usage is sometimes more than mere taste, judgment
and education - sometimes it's sheer luck, like getting across the
street."
- E.B. White

The importance of your title cannot be over-stated. Your
title is the sign-board directing readers of a journal - not to
mention readers of abstracting and indexing publications - to
your work. Your title will be of vital importance to secondary
publications in their efforts to catalogue your work correctly, so
that others can most easily find it.

Your title must completely convey the essence of your
project, but do so in the briefest, most succinct manner possible.

Some rules to go by:

* Avoid useless words and phrases such as *An* or *The*, or
Observations on, Studies on, etc.

* Be specific. Bad title: "Cancer and Obesity." Good title:
"Impact of Morbid Obesity on Cancer Risk."

* Avoid "sentence" titles. Bad title: "Obesity proves a vital
factor in the increase of cancer risk."

* Abbreviations be damned. They tend to screw up indexing in secondary sources. Bad: PCBs. Good: Polychlorinated Byphenyls.

* Avoid numbered series titles. Perhaps not every paper in a series will be published, leaving blanks, for example, between item 10 and item 12. Also series headers tend to be vague and useless. Bad: "Studies in Cancer Prediction. V. Impact of Morbid Obesity on Cancer Risk." The only, important, information-bearing aspect of this is the second phrase. Go with that.

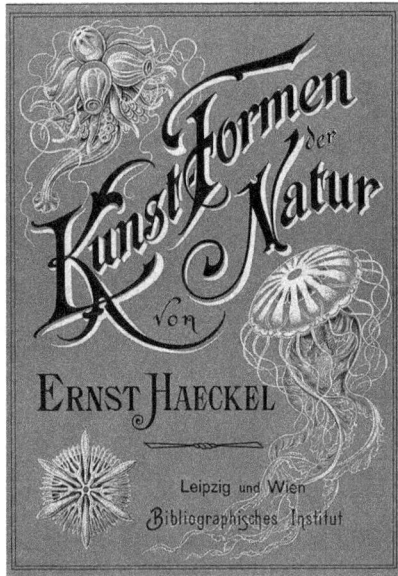

EGO TYRANNOUS: THE MOVIE CREDITS

"Vanity is my favorite sin."

- Al Pacino

Hopefully all authors of a research paper are in agreement about the hierarchy of those who have contributed to the project. Assuming good will among workers, this is the easy part of establishing credit-rank.

The hard part is sifting through the chaos of protocol - or rather lack of protocol - to translate your mutually agreed ranks into an order for publication.

In some journals, rank simply does not apply as the editors insist on an alphabetical listing of authors. This is probably the best - and certainly simplest - approach. I, for one, would love it if the protocol were adopted everywhere. But right now, alphabetical ranking remains primarily a European practice, with American approaches being far more random.

Generally in the United States the first listing is reserved for the senior worker in the research described, regardless of that author's rank in the scholarly community. On a specific project, a graduate student might well outrank a Nobel Prize winner, depending on who came up with the idea for the study in the first place, and who led the way in conceptualizing and executing methodology.

Some institutions impose guidelines whereby the heads of labs are always listed as authors, albeit last authors, on any research conducted in that lab, regardless of whether or not the head had any direct role in the project. Other institutions will insist that virtually every worker in a given lab be listed, in an excess of democracy, whether or not they had any role whatsoever. Regardless, all authors listed take intellectual responsibility for the validity of the research reported.

So, expect the protocol of listing authors to very much depend upon where you do your work - that and the practice of the journal in which you are publishing.

As regards the form of names, please be kind to the indexing services. Use full names including middle initial. Authors who use only their first two initials and last name create vast database confusion for retrieval services and *will* inhibit the proper finding and distribution of your work, as well as the awarding of the credit for your work within your professional community.

Few scientific journals list degrees following names. Most medical journals do, either on the masthead of the articles or in footnotes. For no discipline, however, are degrees or such titles as "Dr." used in Literature Cited.

Listing Institutions and Addresses

"All institutions are prone to corruption and to the vices of
their members."

- Morris West

Here the protocols are simple.

One author? List the address of the lab in which the experiment was conducted. If that author has since moved to another institution, put the new address/affiliation in a footnote.

Two or more authors? All from the same institution? Then list the address once for all authors. Authors all from different institutions? List the institutions in the same order as are the authors.

Five authors from four institutions, two being from the same place? Indicate affiliations with a superscript a, b, c, etc. after each author's name and before the applicable institution/address.

E. GUNDLACH.

MICROSCOPES.

No. 182,919.

Patented Oct. 3, 1876.

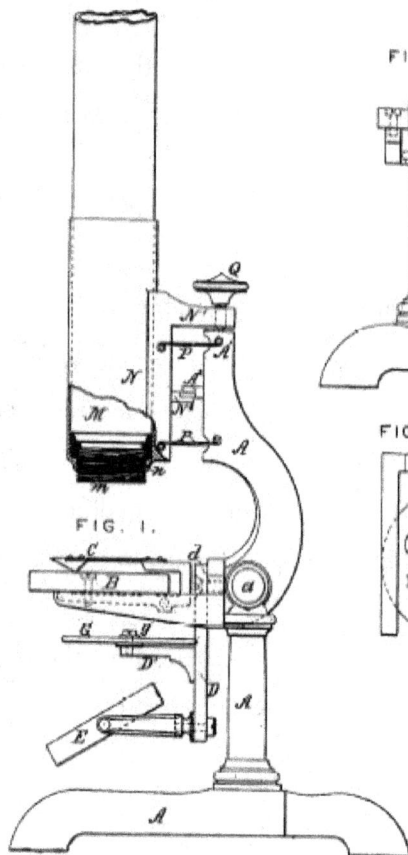

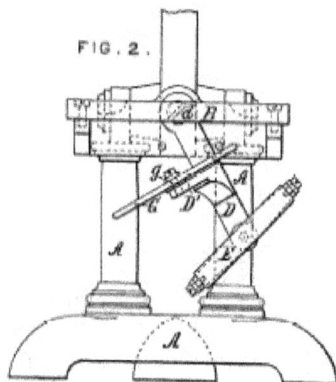

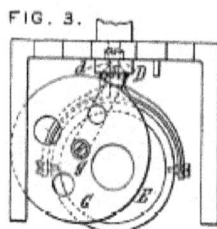

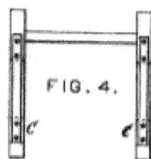

FIG. 2.

FIG. 3.

FIG. I.

FIG. 4.

WITNESSES:

INVENTOR.

22

In Short: The All-Important Abstract

"I believe more in the scissors than I do in the pencil."

- Truman Capote

There is nothing so hard to do as summarize, condense, encapsulate. Nor is there anything so important to do as summarize, condense, encapsulate - at least not when it comes to scientific research papers.

According to the American National Standards Institute (ANSI), a correctly-done abstract "enables readers to identify the basic content of a document quickly and accurately, to determine its relevance to their interests, and thus to decide whether they need to read the document in its entirety." Your abstract is your billboard shouting to passing motorists to slow down and consider what you have to offer. It is also the first thing that journal editors will look at when it comes to your paper.

No properly-prepared abstract ever exceeds 250 words. 200 words are better. And 100 words better still. As well, almost no properly-prepared abstract should ever be more than one paragraph, although occasionally in medical journals abstracts may be divided into several diminutive paragraphs.

Within that tiny space of 250 words (or less), your abstract must condense the primary objectives and scope of the experiment and/or investigation being discussed, the methods employed, the results, and the conclusions. Per C.J. McGirr: "When writing the abstract, remember that it will be published by itself, and should be self-contained. That is, it should contain no bibliographic, figure, or table references. ... The language should be familiar to the potential reader. Omit obscure abbreviations and acronyms. Write the paper before you write the abstract, if at all possible."

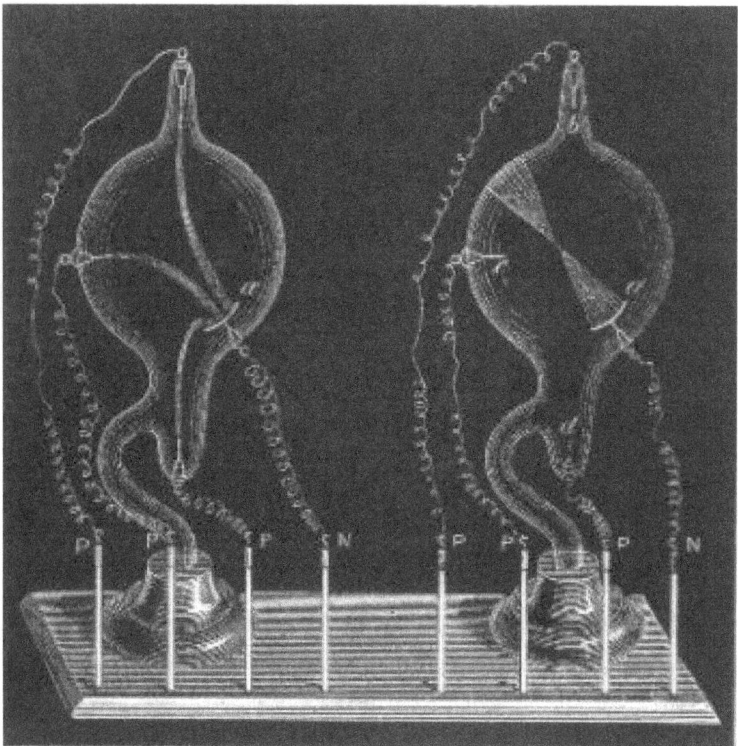

HOW TO BEGIN

"The beginning is the most important part of the work."

- Plato

Don't be misguided by the sequence of early chapters in this book. Your title and abstract ought to be - in fact, *must be* - composed only *after* your paper has been finalized. Otherwise it will be impossible for you to compress and condense your key points accurately and in the most informed, disciplined manner possible.

This being said, it is of course nevertheless important for you to have complete command of your facts and main conclusions before you touch fingers to keyboard. You will naturally be working from notes kept during the experimentation process, which is your logical first resource. But many find, as they formalize their thoughts for final publication, that new observations arise from this unique process - observations born from the clarity of well-wrought composition. Welcome these, and recognize that the act of writing your research paper is itself a part of your research, albeit the very last part. It is not a secondary act, but rather a tertiary act.

Your Introduction

Your Introduction explains to your reader the background to your research. The "job" of your Introduction is to provide enough detail to your reader so that he or she can judge the efficacy, accuracy and relevancy of your work without having to go back and refer directly to documentation for previous studies. In addition to orienting the reader with a highly condensed summary of previous research, your Introduction must also explain your precise purpose in conducting your own research, *summarize* the methodology of your investigation and, where necessary, your reason(s) for choosing such methodology, elucidate your key results, and enunciate the key conclusions you have derived from these results.

Introductory Housekeeping

Have you previously published anything related to the research described in your paper, such as a note or abstract? Be sure to cite this in your Introduction. Also cite any papers related to your own research which have very recently or soon will be published, with citations. You do this as a courtesy to your readers, to make it as easy possible for them to keep abreast of all the most immediately relevant work in your particular area of inquiry.

For Whom are You Writing?
And What Particular Needs Might They Have vis-à-vis
Introductory Information?

Will your paper be read primarily by those with a fairly advanced and detailed knowledge of your unique area of work? Or will it be read by people who, though they work in related disciplines/avenues of research, nevertheless do not have an intimate knowledge of the subject at hand? If the latter, then your Introduction should - in addition to the tasks described above, and as a courtesy to these readers - make a point of defining discipline-specific abbreviations, terms and jargon contained within your presentation. This will make their reading easier, and eliminate any potential confusion in their understanding of your information/analysis. In short, everybody wins.

JOANNIS KEPPLERI
Sac. Cæf. Majeft. Mathematici

DE

STELLA NOVA

IN PEDE SERPENTARII, ET
QUI SUB EJUS EXORTUM DE
NOVO INIIT,

TRIGONO IGNEO.

LIBELLUS ASTRONOMICIS, PHYSICIS, META-
physicis, Meteorologicis & Astrologicis Disputationibus,
εὐδόξοις & παραδόξοις plenus.

ACCESSERUNT

I. *DE STELLA INCOGNITA CYGNI:*
Narratio Astronomica.

II. *DE JESV CHRISTI SERVATORIS VERO
Anno Natalitio*, consideratio novissimæ sententiæ LAV-
RENTII SVSLIGÆ Poloni, quatuor annos in usitata
Epocha desiderantis.

Cum Privilegio S. C. Majeft. ad annos XV.

GRANA DAT E FIMO SCRUTANS

PRAGAE
Ex Officina calcographica PAULI SESSII.
ANNO M. DC.VI.

MATERIALS, METHODS AND OTHER THINGS THAT GO BUMP IN THE NIGHT

"But man has still another powerful resource: natural science
with its strictly objective methods."

- Ivan Pavlov

Your Introduction included an encapsulated digest of your methodology. In the formal section of your paper elaborating methodology, you must expand upon this, providing all the details necessary for any reader of your paper to reproduce your results in his or her own laboratory. To this end, carefully describe and define every aspect of your experimental design.

Materials

With regard to materials, you must include precise descriptions of technical specs, quantities, and method(s) of preparation. Trade names should be avoided wherever possible; endeavor to use chemical and/or generic names. In an instance where you must refer to a material by a trade name, capitalize this and then use a generic term immediately thereafter: CHARMAN, paper towels.

Any organisms, plants or animals used must be identified in the greatest detail possible, down to genus, species, and strains. List all sources. Note all special characteristics (age, sex, etc.). As regards human subjects, provide complete information on your method of selection, and include "informed consent" statements as required.

Methods

Keep this portion chronological: first assay, second assay, and so forth - in building-block order. But be sure to include the methodology for each and every assay involved in the project. In this and other sections of your Materials and Methods presentations, use subheadings as necessary. Ideally, the subheadings here should align with the same subheadings used in Results, although this is not always possible.

Precision in All Things, Particularly Measurement

No detail is too minute. Don't just describe *heated water*, but rather *distilled water heated to 372 degrees Fahrenheit*. Heated how, and in what quantity? Answer these questions. With regard to your statistics, focus on data rather than statistical approaches; only discuss particular statistical methods if these vary in any way from the norm. Leave no fragment of procedure unstated. Explicitly state materials, distillation temperatures, apparatus, etc.

References

References can, at times, save you some work. Has your methodology been used previously by others? Refer to them, and save yourself some ink. With a newly improvised/propounded method you will, of course, have to describe all methodology in grim detail, thus allowing others to refer to you down the road.

Avoid Allusion to Results

It is very tempting, in elucidating Methods and Materials, to allude also to Results. Do not do this. Stop at procedures, apparatus and methods.

NICOLAI
COPERNICI TO-
RINENSIS DE REVOLVTIONI-
bus orbium cœlestium,

Libri VI.

IN QVIBVS STELLARVM ET FI-
XARVM ET ERRATICARVM MOTVS, EX VETE-
ribus atq́ recentibus obseruationibus, restituit hic autor.
Praeterea tabulas expeditas luculentasq́ addidit, ex qui-
bus eosdem motus ad quoduis tempus Mathe-
matum studiosus facillime calcu-
lare poterit.

ITEM, DE LIBRIS REVOLVTIONVM NICOLAI
Copernici Narratio prima, per M. Georgium Ioachi-
mum Rheticum ad D. Ioan. Schone-
rum scripta.

Cum Gratia & Priuilegio Cæs. Maiest.

BASILEAE, EX OFFICINA
HENRICPETRINA

PRESENTING RESULTS

"Realists do not fear the results of their study."

- Fyodor Dostoevsky

Generally, your Results section is comprised of two key components. The first describes your experiments in broad strokes, providing a wide-angle-lens view of things while not reiterating the details you've already elucidated in Materials and Methods. The second supplies the data which you've produced - though not necessarily all the data. Spare your reader (and your reputation) by summarizing results and selecting those aspects of results which relate most directly to your Discussion, which will follow.

Throughout your paper, brevity equals clarity, both for you and your reader. Condense, condense, condense. Cite *meaningful, necessary* data while dispensing with data unnecessary to your main focus. Remember the words of S. Aaronson, who said: "The compulsion to include everything, leaving nothing out, does not prove that one has unlimited information; it proves that one lacks discrimination."

Say that in the course of your work you experimented with a large number of variables. A few of these variables affected the matter/behavior being studied, and a great many didn't. Variables of the former group need be related/presented in detail; but those of the latter group may be disposed of with one sentence, this attached to a footnote itemizing the variables or group/class of variables that had no impact.

If your research has yielded only a few essential determinations, elucidate these within the text of your Results section. Numerous, repetitive determinations should, however, be related via tables and graphs. (Best practices re: tables and graphs are elucidated later in this volume.)

Most importantly, avoid the "redundancy trap" into which so many researchers fall when composing their Results presentation. Do not restate the obvious. Do not feel required to enunciate facts already made clear to the reader in data otherwise related - such as repeating information in text that has already been conveyed via a graph. Also, here as elsewhere, avoid unnecessary verbiage: over-embroidered language. Bad: "As decisively demonstrated in tables 1 and 2, toxins at the second level tend to be self-inhibiting ..." Good: "Toxins at the second level tend to be self-inhibiting ..."

How to Write Your Discussion

"A man who uses a great many words to express his meaning is like a bad marksman who, instead of aiming a single stone at an object, takes up a handful and throws at it in hopes he may hit."

- Samuel Johnson

Your Discussion is the most important part of your paper. As well, it is almost always the toughest to compose. In your previous sections, you have revealed vital, original Results and shown precisely how you've come up with that data. Now you have to explain what, in your opinion, these Results mean.

Do not reiterate your Results. Instead elucidate key generalizations, principles and relationships indicated by the Results.

Explain the logic that allows you to accept or reject your original hypotheses.

Make clear where and how your Results vary from, confirm or refute Results published by other researchers, aka: existing theory and knowledge.

Engage in "full disclosure" regarding any potential glitches in your data - otherwise count on vigorous readers (or worse, journal editors) to do so.

Enunciate all conclusions concisely and clearly, and summarize all your evidence for each conclusion.

Speculate as necessary but be sure to identify each speculation as such.

And blow your horn. Explain the implications of your Results - what they say about the path of future work, whatever practical applications they may indicate, and how they fit into the broad scheme of research on the given topic.

The last point can't be emphasized enough. In the closing of your Discussion, you absolutely must make clear the *significance* of your Results - why the reader should *care*.

How to Write Your
Acknowledgments

"Even for the physicist, the description in plain language will
be a criterion of the degree of understanding that has been
reached."

- Werner Heisenberg (Nobel Prize winner in physics)

Here, quite simply, you will acknowledge anyone not a co-author who leant material or intellectual support to the study at hand. This would include any colleagues with whom you discussed the basis and design of your experiment before you carried it out, anyone who leant or gave you required materials, and also outside reviewers who commented on the preliminary drafts of your paper. You must as well note any and all sources of funding supporting your research. Keep this section short and business-like, verging on the perfunctory.

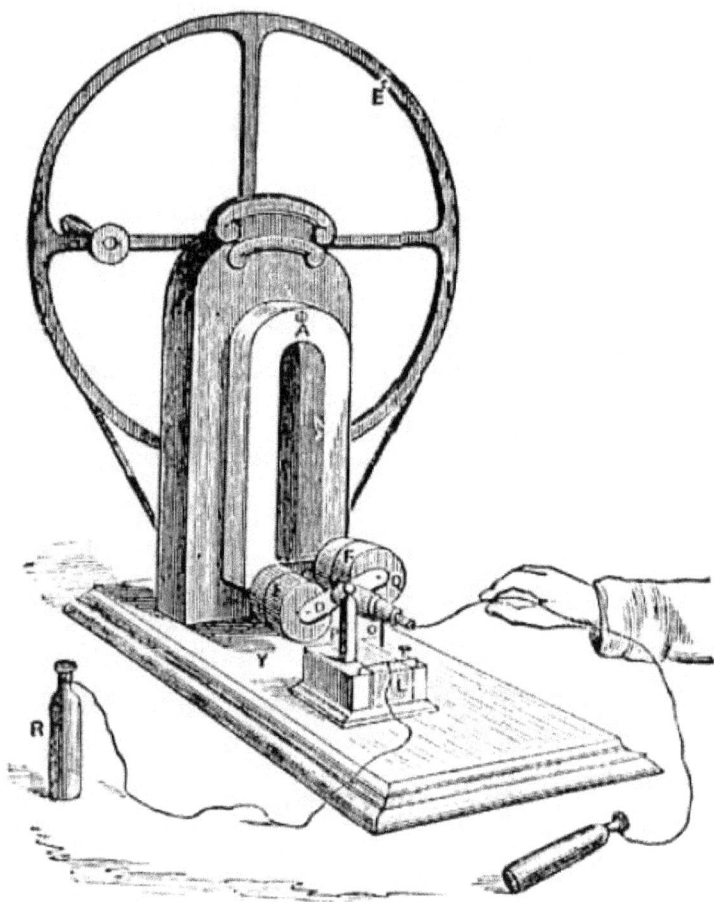

LITERATURE CITED

"Nature composes some of her loveliest poems for the
microscope and the telescope."
- *Theodore Roszak*

Your Literature Cited section follows your discussion, and
wraps up your paper. The most standard format for this is as
shown in the following example:

Doyle, M., D.P. Clancy and I.M. Jones. 1987. Excalabrosis: Refined
remnants of chordivastas. Annual Review of Protoploptology 25:
145-168.
Lappin, M. J. and C.U. Wu. 1963. Divided organisms of mole
droppings. Mole Physiology 70:101-152.
Moran, S.A. and J.O. Tweed. 1967. The Theory of Mole
Husbandry. Yale University Press, New Haven, CT.
Poe, E.A. 1949. Raven and mole cross-contamination. American
Journal of Bestiary 35:1066-1071.

Key style elements:

- Alphabetized list.

- No first or middle names.

- Only the first words in the titles of journal articles (except for proper nouns) are capitalized.

Using Tables

"The theory of probabilities is at bottom nothing but
common sense reduced to calculus."

- Laplace, *Théorie analytique des probabilités*, 1820

What the definition of *ironic*? How about this: I've already
stated *more than once* that it is important *never to be unnecessarily
repetitive* in the presentation of Results. Sometimes, however, a
certain measure of repetition is vital to making or enforcing a key
point or points. In such instances, a table is often the most
efficient and compact means for conveying such data.

As previously mentioned, if you have relatively few
determinations to present, these need only be elucidated within
the text. But generally, when presenting five or more
determinations, it is worthwhile to consider a tabular approach.

A few key guidelines for constructing optimal tables:

* Of course, be sure to number your table for ready-reference
within the text.

* Also, of course be sure to title your table, describing its
contents as accurately and concisely as possible.

* Arrange your table so that information flows vertically
rather than horizontally. Have like elements read down rather

than across, while comparative data should read across. This latter rule applies especially to any numerical data, which should always be shown in columns rather than rows.

To Graph or Not to Graph

"Everything should be made as simple as possible, but not simpler."

- Einstein

A table is a table. A graph is a table rendered as an illustration. And simplicity is the main virtue to be sought in both.

As Edward Tufte wrote in his now-classic book *The Visual Display of Quantitative Information*: "Excellence in statistical graphics consists of complex ideas communicated with clarity, precision, and efficiency. ... What is to be sought in designs for the display of information is the clear portrayal of complexity. Not the complication of the simple; rather the task of the designer is to give visual access to the subtle and the difficult - that is, the revelation of the complex."

When should a graph be used, instead of a table? Graphs are best for the presentation of trending data. A graph with all straight lines will be dull, providing information that might as well be presented as straight numerical data in a table. On the other hand, a graph with curved, nuanced lines conveys subtleties in data that would not be easily discerned if confined to a table. So, let the nature of your data dictate your approach.

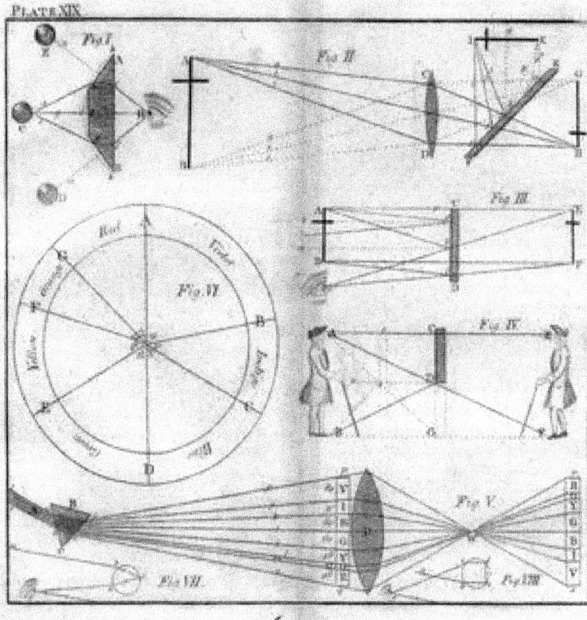

15

The Art of the Review Paper

"To raise new questions, new possibilities, to regard old problems from a new angle, requires creative imagination and marks real advance in science."

- Einstein

Review papers do not present original research. Instead, they survey the previously published literature and put it into context. But the job is not to merely summarize and/or condense. No, the job here is to evaluate the relative merits of the various publications under consideration, and elucidate conclusions derived from this evaluation. (Note: Review paper content requirements will vary from one journal to the next. While most journals emphasize the critical evaluation of publications, a few are more interested in general summaries of the literature. The most vital review papers are the former.)

Generally, the IMRAD format does not apply here, although one might incorporate a Methods discussion which explains any precise methodology of review or analysis being employed. Generally, however, the format of a review paper is Introduction followed by Discussion followed by Conclusion.

As with all writing - not just scientific writing - it is important to understand the profile of your intended audience before beginning. Audiences for review papers tend to be less highly specialized, and to have a less intimate understanding of the subject at hand, than audiences for peer-reviewed accounts of primary research. The range of papers reviewed will interest a range of readers, including worker-researchers in related disciplines and sub-fields. Given this, you must assume less specific knowledge on the part of your reader base. Do not assume that your readers will understand discipline-specific jargon or abbreviations. If you absolutely have to use these, be sure to define them.

Your greatest challenge, given the reader-base, is the summary explanation of difficult concepts for the benefit of those readers who need such summaries. These summaries must, of course, be meaningful. But they must also be brief. Meaningful and brief - a tough combination of requirements - but the task must be met.

16

THE ART OF THE CONFERENCE REPORT

"There are many hypotheses in science which are wrong.
That's perfectly all right; they're the aperture to finding out
what's right. Science is a self-correcting process. To be accepted,
new ideas must survive the most rigorous standards of evidence
and scrutiny."

- Carl Sagan

Most conference reports are not considered, nor meant to be considered, valid primary publications. Conference reports are, for the most part, basically just collections of review papers or else they contain information of a speculative nature, with scientists pondering results which they may not yet be confident enough of to offer for primary publication. As well, the majority of conference reports are not arbitrated by peer review.

For papers included in those conference reports adjudged not to be primary publications, you need not provide data required for the reproducibility of results. In other words, you can generally skip providing Materials and Methods information.

You can also skip the typical review of pre-existing literature.

In your conference report paper, which should usually run between 1,000 and 2,000 words, you can and should simply provide your recent results and then speculate as to what those results mean. This is all preliminary. Save all details of how you arrived at your results for later primary publication.

It is here - as opposed to your later valid primary publication - where you can give the freest reign to your speculations re: possible alternative answers and options for further experimentation/inquiry. (Journal editors will give you much less latitude when it comes to speculation as presented in your formal presentation of results.)

The broad outline of a typical conference report is as follows: First, briefly, define the problem and/or question being addressed. Second, briefly describe results (with no more than three tables or illustrations). And last, provide your speculations - not as briefly.

17

THE ART OF THE SCIENTIFIC BOOK REVIEW

"I love criticism just so long as it's unqualified praise."
- Noel Coward

There are essentially four different types of science books: monographs, reference works, textbooks and popular trade books. Each category has its own unique character, and each demands unique approaches when it comes to criticism.

Monographs tend to be works written by scientists for scientists. They are pitched at a very high level of technical expertise, and assume much knowledge on the part of the reader. Most monographs consist of many authors, each tackling a chapter, but on some occasions a monograph will have a single author.

When reviewing monographs, it is essential to place the work in the context of the previous published literature, and enunciate exactly how the work extends and enhances that literature. There should be no quibbling about prose, as their might be in a review of a general trade book. Rather, critical emphasis should be on the range and quality of the work presented, and the skill with which it has been integrated into a single, vital volume.

Does the monograph truly represent the state of the art, or has it already been superseded before even coming to press? (This happens, especially in those fields where research runs at a fast and furious pace.) Does the monograph present diverse possibilities and conclusions, giving a free reign to reasonable speculation, without adhering to the "party line" of any one particular school of research or thought? Are the credentials of those contributing to the monograph up to snuff? If a part of a series (*Advances in ...*, *Annual Review of ...*), does the work maintain the standard of excellence set by previous volumes, and does it logically proceed from the research narrative as left off in the previous number?

These are all questions which must be addressed in a review of a scientific monograph. Other types of works, meanwhile, demand different critical approaches.

Reference books condense and catalog various types of scientific data. We speak here of discipline-specific dictionaries, bibliographies and encyclopedias. When reviewing these, the key questions are reliability/timeliness of data, and ease of access (as in, the intuitive logic by which the information is arranged). Also worth considering: the authority of the editor and/or organization assembling the data. When writing a review of a reference work, your main audience is potential users of the work and, secondarily, staff at libraries and laboratories who might be considering purchase and/or subscription to the work.

Textbooks are, of course, volumes used in the teaching of courses. When reviewing texts, the chief concerns to be addressed are how clearly the prose of the volume explains complex topics, how effectively the author or authors use illustrations and/or tab data to convey essential information, and whether the information provided represents the very latest accepted knowledge. When writing a review of a textbook, keep in mind that your primary audience is professors and other instructors, as well as those department heads who will make adoption decisions based upon what you say.

The textbook market is poised for radical change in the near future. For many years, both professors and students have been discomforted by the egregious prices charged for standard texts published by such houses as McGraw-Hill, Norton, Wiley, Harcourt and Houghton Mifflin. Digital publishing combined with the ubiquitous nature of the Internet have recently enabled a number of "open" textbook initiatives which not only bring costs down, but also allow instructors to customize texts to conform to their precise syllabi, thus (hopefully) enhancing the classroom experience overall.

The Community College Open Textbook Collaborative, Currika (K-12 Open Curricula Community), and Flat World Knowledge are a few of the firms spearheading a radical shift in the way textbooks are created, customized, updated and distributed. As this fluid situation evolves, the challenge for reviewers will be to curate texts emerging from an entirely new process of collaboration, and assure quality through steadfast, fair commentary which sets a benchmark of excellence.

Trade science books are those distributed en masse through trade bookstores (those few that are left) and, increasingly, via e-editions for the Kindle, Nook, iPad and other such e-Readers and tablets. These volumes - Stephen Hawking's *A Brief History of Time* would be a good example - are written for a popular audience and thus assume no in-depth knowledge of any given scientific discipline. So too must your review make the same assumption, clarifying terms and jargon for potential readers of the work, and also making the case for (or against) the work as a worthy introduction to the field. In such reviews as this, it is fine to comment not only on informational accuracy, but also on the smoothness of prose (the artfulness of the writing), as trade books are meant, first and foremost, for pleasure, even if it is the pleasure of learning.

Regardless of what category of work you are reviewing, always keep in mind that *your charter is to discuss the book, and the book's treatment of its topic,* rather than the topic itself. If you want to do the latter, write a book.

Plagiarism

"For such kind of borrowing as this, if it be not bettered by
the borrower, among good authors is accounted plagiary."
- John Milton

"Copy from one, it's plagiarism; copy from two, it's research."
- Wilson Mizner (Playwright)

We all remember the big dust-up several years ago when
noted presidential historian Doris Kearns Goodwin was revealed
to have plagiarized significant portions of her book *The
Fitzgeralds and the Kennedys*. Countless other scandals across a
range of disciplines come to mind. In 2008, well known British
psychiatrist Raj Persaud found himself suspended from practicing
for three months after confessing to having plagiarized several
other authors in articles he published in medical journals and
newspapers.

Of course, there are varying levels and degrees of plagiarism,
ranging from the inadvertent to the premeditated. Even more
importantly, in the context of scientific research papers, is the
nature of the information being re-used, and its prominence

and/or importance within the context of the offending piece of literature.

Writing in 1984, Brian Martin (Professor of Social Sciences at the University of Wollongong, Australia) said: "The significance of plagiarism can vary widely, depending on its extent, strategic location, and the context in which it occurs. An isolated instance of plagiarism - one sentence or paragraph, for example - would not usually be cause for concern, whereas a paper copied almost *verbatim* would be considered a gross violation of academic norms. Strategic location refers to centrality in an academic presentation. Plagiarism in crucial points of argumentation is more serious than in a largely extraneous literature review. Finally, the overall context of plagiarism must be considered: the nature of the contribution, scholarly or otherwise."

Writing in *Nature* in 2008, Mounir Errami and Harold Garner of The University of Texas Southwestern Medical Center commented: "Scientific productivity, as measured by scholarly publication rates, is at an all-time high. However, high-profile cases of scientific misconduct remind us that not all those publications are to be trusted - but how many and which papers? Given the pressure to publish, it is important to be aware of the ways in which community standards can be subverted. ... There are legitimate and illegitimate reasons for two scientific articles to share unusual levels of similarity. Some forms of repeated publication are not only ethical, but valuable to the scientific community, such as clinical-trial updates, conference proceedings and errata. The most unethical practices involve substantial reproduction of another study (bringing no novelty to the scientific community) without proper acknowledgement. If such duplicates have different authors, then they may be guilty of plagiarism, whereas papers with overlapping authors may represent self-plagiarism. Simultaneous submission of duplicate articles by the same authors to different journals also violates journal policies."

The best definition of self-plagiarism that I've found comes from Wikipedia: "*Self-plagiarism* (also known as 'recycling fraud') is the reuse of significant, identical, or nearly identical portions of one's own work without acknowledging that one is doing so or without citing the original work. Articles of this nature are often referred to as duplicate or multiple. In addition to the ethical issue, this can be illegal if copyright of the prior work has been transferred to another entity. Typically, self-plagiarism is only considered to be a serious ethical issue in settings where a publication is asserted to consist of new material, such as in academic publishing or educational assignments."

But even then, self-plagiarism remains something of a gray area. "Sometimes [text reuse] is just unavoidable," Catriona Fennell, director of journal services at Elsevier, told Jef Akst of *The Scientist.* "Really, how many different ways can you say the same thing?" Blogging on *The Scientist* web site in September of 2010, Akst added: "Because scientists tend to study the same topic over many years or even their entire careers, some aspects of their research papers, particularly the literature review and methodology, will be repeated. Once they've figured out how to word it succinctly and accurately, some argue, it's best left unchanged. 'You're laying the groundwork for an ongoing discussion [so] making changes might actually be a bad idea,' [Patrick] Scanlon [of the Rochester Institute of Technology's Department of Communication] said. 'It would muddy the waters.'"

Generally, where conscious self-plagiarism is involved in your paper, you are best advised to flag that self-plagiarism up-front for journal editors, and see what their take is on your usage of your previous published material. As has been indicated above, in some instances this will not be considered a problem, while in other cases it will.

You will have noticed, in the paragraphs above, a great many words composed and spoken by a number of people other than the writer of this book. How is this not plagiarism? The answer is simple. *Attribution.* So long as sources and authors are clearly identified either within text or through footnotes and/or endnotes, and their words used within quotes, plagiarism is not at issue. Plagiarism only arises when one individual claims the words, thoughts and data of another as his/her own original work or revelation.

Plagiarism has never been a winning game. Through the centuries plagiarists have left ticking time bombs behind them in the form of their own publications. The first witness against all plagiarists is their body of work, and the evidence of theft to be found there if not today, then tomorrow. In recent times, digital tools have made it easier than ever for plagiarism to be discovered and tracked. A host of plagiarism detector tools on the Internet use complex algorithmic searches to document purloined phrases and even close paraphrases (word-substitution plagiarism). Most journal editors now use at least one of these tools to conduct originality-checks on submitted prose.

To commit plagiarism is to engage in professional suicide. Nothing will kill a career, or a reputation, faster than a charge of intellectual theft. And for students at universities, plagiarism is usually grounds for expulsion. So, in less than three words: *steer clear.*

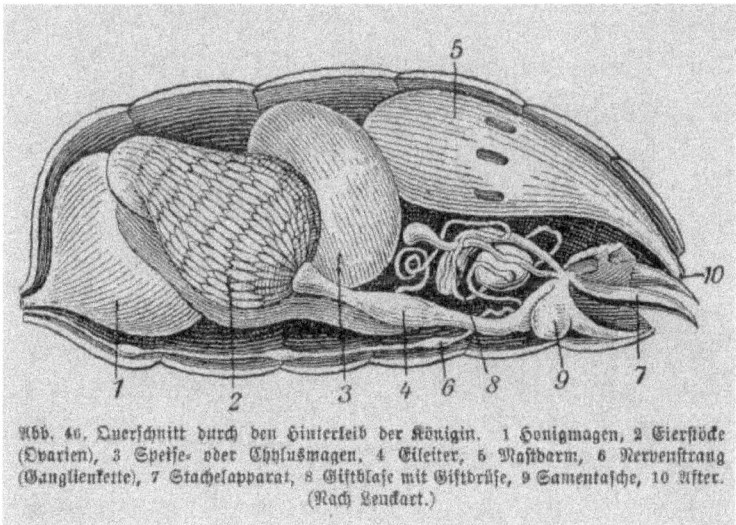

Abb. 46. Querschnitt durch den Hinterleib der Königin. 1 Honigmagen, 2 Eierstöcke (Ovarien), 3 Speise- oder Chylusmagen, 4 Eileiter, 5 Mastdarm, 6 Nervenstrang (Ganglienkette), 7 Stachelapparat, 8 Giftblase mit Giftdrüse, 9 Samentasche, 10 After. (Nach Lenckart.)

IL SAGGIATORE

Nel quale
Con bilancia esquisita e giusta
si ponderano le cose contenute
nella

LIBRA ASTRONOMICA E FILOSOFICA
DI LOTARIO SARSI SIGENSANO

Scritto in forma di lettera

Al*l*mo & Reueren.mo Monsig.re D.

VIRGINIO CESARINI

Acc.° Linceo M.° di Camera di N.S.

Da

GALILEO GALILEI

Acc.° Linceo Nobile Fiorentino
Filosofo e Matematico Primario
del
Ser.mo Gran Duca di Toscana.

FILOSOFIA
NATVRALE

MATEMATICA

IN ROMA M.DC.XXIII.
Appresso Giacomo Mascardi

F. Villamoena Fecit.

19

HOW TO SUBMIT YOUR WORK

"Science is built up of facts, as a house is built of stones; but an accumulation of facts is no more a science than a heap of stones is a house."

- Jules Henri Poincaré, mathematician, physicist and philosopher

In addition to being published simultaneously on the web and in print, these days most scientific journals offer web based submission engines through which they insist that you submit your paper. Within the same web presence you will invariably find "Instructions for Authors" giving details on formats, word counts, and other specifics as prescribed by the editors of the specific publication at hand. It is important that you painstakingly adhere to these, or risk having your work dismissed out-of-hand.

20

A Few Fundamental Rules to Improve Your Writing

"I try to leave out the parts that people skip."

- Elmore Leonard

There are a few basic rules which, if applied correctly, will go a long way toward improving the flow and comprehensibility of just about anyone's writing. Here they are:

1) Use the passive voice sparingly. An active voice keeps your text flowing along, a passive voice slows it down. *Wrong:* "Only at locotistic sites was the co-localization of LME2 and NF3 with the PSL-95 complex observed." *Right:* "We observed that LME2 and NF3 co-localize with the PSL-95 complex only at locotistic sites."

2) Use propositional phrases (such as noun strings, nominalizations [noun forms of verbs] and the verb "to be") sparingly, in order to avoid confusion. Too many of these phrases equal cognitive chaos.

3) Brevity is best. As Strunk and White put it in *The Elements of Style*: "Vigorous writing is concise. A sentence should contain no unnecessary words, a paragraph no unnecessary sentences, for the same reason that a drawing should have no unnecessary lines and a machine no unnecessary parts. This requires not that the writer make all his sentences short, or that he avoid all detail and treat his subjects only in outline, but that every word tell."

4) Emphasize nouns and verbs over adjectives and adverbs. The former tend toward objective language, the latter toward subjective.

5) Never say in 20 words what you can say in 10.

6) Avoid tired, unoriginal phrases (aka, *clichés*).

7) Respect the logic of grammar. Especially, make sure that the subject of a sentence really does what the verb says it does. Saying "Researchers must take gross samples using the X instruction" (incorrect) is not the same as saying "Researchers must use the X instruction when they take gross samples" (correct).

8) Favor simple descriptions.

9) Favor short, simple words. Use *learn* instead of *ascertain*, *try* instead of *endeavor*, *list* instead of *enumerate*, *start* instead of *initiate*, *change* instead of *modify*, *use* instead of *utilize*, and so forth.

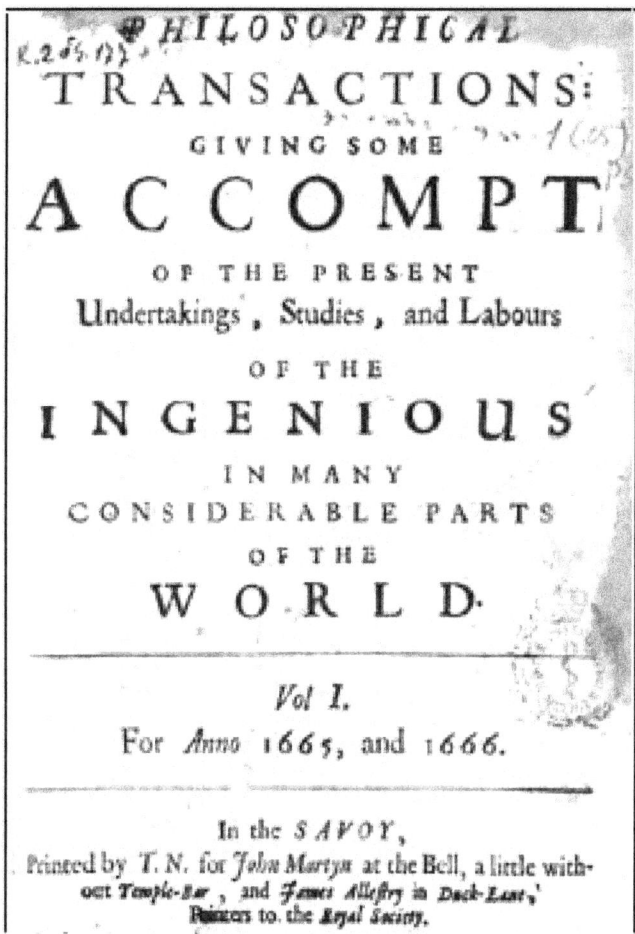

PHILOSOPHICAL
TRANSACTIONS:
GIVING SOME
ACCOMPT
OF THE PRESENT
Undertakings, Studies, and Labours
OF THE
INGENIOUS
IN MANY
CONSIDERABLE PARTS
OF THE
WORLD.

Vol I.

For *Anno* 1665, and 1666.

In the *SAVOY*,
Printed by *T. N.* for *John Martyn* at the Bell, a little with-
out *Temple-Bar*, and *James Allestry* in *Duck-Lane*,
Printers to the *Royal Society*.

Prodromus

DISSERTATIONVM COSMOGRAPHICARVM, CONTINENS MYSTERIVM COSMOGRAPHICVM,

DE ADMIRABILI

PROPORTIONE ORBIVM COELESTIVM, DEQVE CAVSIS

cœlorum numeri, magnitudinis, motuumque periodicorum genuinis & proprijs,

DEMONSTRATVM, PER QVINQVE
regularia corpora Geometrica,

A

M. IOANNE KEPLERO, VVIRTEMbergico, Illustrium Styriæ prouincialium Mathematico.

Quotidiè morior, fateorque: sed inter Olympi
Dum tenet assiduas me mea cura vias:
Non pedibus terram contingo: sed ante Tonantem
Nectare, diuina pascor & ambrosiâ.

Addita est erudita NARRATIO M. GEORGII IOACHIMI RHETICI, de Libris Reuolutionum, atq, admirandis de numero, ordine, & distantijs Sphararum Mundi hypothesibus, excellentissimi Mathematici, totiusq, Astronomiæ Restauratoris D. NICOLAI COPERNICI.

TVBINGÆ
Excudebat Georgius Gruppenbachius,
ANNO M. D. XCVI.

21

Common Grammar Traps

"Grammar is a piano I play by ear. All I know about
grammar is its power."

- Joan Didion

Here are the key aspects of grammar that most often trip
people up:

Avoid splitting infinitives, which means avoid placing a word
between *to* and the verb following. That is, avoid: *to rapidly walk,
to boisterously sing.* Correct: *to walk rapidly, to sing boisterously.*

Avoid unrelated participles. Wrong: *Walking up the street a
dog bit me.* Right: *Walking up the street I was bit by a dog.*

Remember that a singular collective noun always calls for a
singular verb. Correct: *A sample of ten strains is to be used.*

Avoid ending sentences with prepositions. Incorrect: *Is there
a library I can find the book in?* Correct: *Is there a library where I
can find the book?*

Remember that the words "either," "neither," "each," "every" and "any" always take singular verbs. Correct: *Each lab received a handbook.*

Always treat "none" as a plural. Correct: *None of the results are satisfactory.*

Treat "kind," "sort," and "type" as singular nouns. Correct: *This type of result is common.*

Use "less" for quantity, and "fewer" for number. Correct: *Less heat should be applied.* Correct: *Fewer than a dozen results were in line with expectations.*

Use "between" for two things, "among" for a group. Correct: *We placed a divider between the two samples.* Correct: *We placed placebos among the samples.*

Make a habit of using "shall" after I/we in statements indicating simple future tense.

Use "will" in statements expressing precise intention.

22

THE PEER REVIEW

"Of all the cants which are canted in this canting world --
though the cant of hypocrites may be the worst -- the cant of
criticism is the most tormenting!"

- Laurence Sterne

Peer review has long been the cornerstone of all serious
scholarly publishing. It has always been a given that one's work
should be rigorously considered and tested by objective
colleagues through either an open or blind (anonymous) peer
review process before publication. Indeed, virtually every serious
scholarly journal identifies itself as either "peer reviewed" or
"refereed" (the same thing.)

Today the peer review remains central to the scholarly
publishing process; but the advent of the Internet and online
publication has caused some to start rethinking the role of peer
reviewing in a number of disciplines. As scholarly publishing has
retooled, so have the potentialities for peer reviewing expanded
and evolved. In fact they have evolved in such a way that some

prominent voices now call for the peer review model to be turned on its head.

Writing in his 2008 book *Here Comes Everybody: The Power of Organizing Without Organizations*, Clay Shirky suggested that the apparatus of the Internet makes it possible to model a "publish, then filter" paradigm - a process allowing for the publication of results with the intention that this be followed by a period of open scholarly criticism, discussion and debate, all online.

Kathleen Fitzpatrick, Professor of Media Studies at Pomona College in Claremont, California, has seconded Shirky's motion, saying that it makes no sense to maintain an editorial process "designed for print's economics of scarcity within the internet's economics of abundance; if what is scarce in the age of the network is not the means of production but the time and attention available for consumption, the best use of peer review would be to help researchers find the right text, of the right authority, at the right time. A born-digital system of review would work with rather than against the strengths and values of the network by privileging the open over the closed, and by understanding the results of peer review as a form of metadata enabling scholars to find and engage with research in their fields."

Issues of credentialing, open vs. anonymous reviews, protocols of community-based filtering, "peer-to-peer" review procedures, preservation (proper tagging and archiving) of drafts and comments, the idea of intellectual property as it relates to the "remixed" results of community peer-review - all these things needs to be thought about carefully and deeply as scholarly publishing pushes forward into the digital age.

But through all of this, one thing has and will remain unchanged: the need for your work to be completely reproducible and for your results and further theorizing to be completely defensible, when tested by the best and brightest of your colleagues.

Note: For most journals, your paper will be reviewed through a *single-blind peer review* (SBPR), in which your reviewer(s) will know who you are, but not the other way around. Occasionally, a *double-blind peer review* (DBPR) will be called for - a process in which you have no word of the identities of your reviewers, and your reviewers have no inkling of your name, credentials or lab location. You will in turn see your reviewer (or referee) comments, and be advised as to the journal's decision on your manuscript. The journal editor can either accept your manuscript as is, reject it outright, or offer to publish with certain changes and/or clarifications urged by the referees.

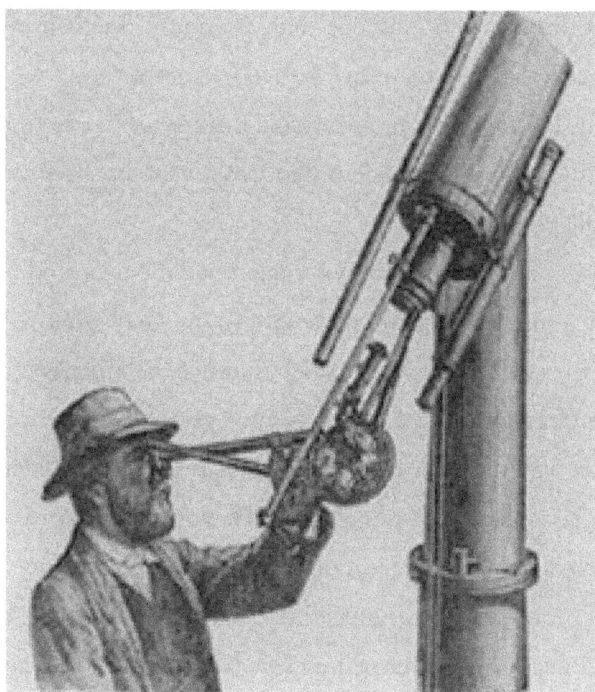

BIBLIOGRAPHY

Ally, Michael. *The Craft of Scientific Writing.* Englewood Cliffs, NJ: Prentice Hall. 1987.

Day, Robert A. *How to Write and Publish a Scientific Paper, 4th Edition.* Phoenix, AZ: Oryx Press. 1994.

Fitzpatrick, Kathleen. *Planned Obsolescence: Publishing, Technology, and the Future of the Academy.* New York: New York University Press. 2011.

Gurak, Laura A. *The Technical Communication Handbook.* New York: Longman. 2008.

Gustavi, Bjorn. *How to Write and Illustrate a Scientific Paper, Second Edition.* Cambridge: Cambridge University Press. 2008.

Iles, Robert L. *Guidebook to Better Medical Writing, Revised Edition.* Iles Publications. 2003.

Katz, Michael J. *From Research to Manuscript: A Guide to Scientific Writing, Second Edition.* New York: Springer. 2009.

Shirky, Clay. *Here Comes Everybody: The Power of Organizing Without Organizations.* New York: Penguin Press. 2008.

Strunk, William and E.B. White. *The Elements of Style, 4th Edition.* New York: Longman. 1994.

Tufte, Edward. *The Visual Display of Quantitative Information.* Cheshire, CT: Graphics Press. 1983.

Zinsser, William. *On Writing Well, 30th Anniversary Edition.* New York: Harper. 2006.

ABOUT THE AUTHOR

Nancy Fox is a leading freelance developmental editor whose clients have included Oxford University Press, Springer-Verlag, Microsoft Press, MIT Press, Addison-Wesley and Yale University Press.

ABOUT THE PUBLISHER

New Street Communications develops, publishes and distributes superior works of nonfiction.

New Street makes books available as print editions and, more importantly, as eBooks - also, through licenses, as audiobooks and in translation.

New Street's focus is on the intersection of digital technology and society; transformative business communication and innovation (particularly the conceptualizing of elegant new tools, markets, products and paradigms); environmental issues; socially-relevant children's literature; travel; and the life and writings of Ernest Hemingway.

New Street's books are authored by distinguished journalists, entrepreneurs, developers and thought leaders.

All New Street eBooks come free of DRM.

Corporate mantra: *Work Smart; Be Kind.*